Richard Honkanen

Virtualization 101 for small business and enterprise

GRIN Verlag

Bibliografische Information der Deutschen Nationalbibliothek:

Die Deutsche Bibliothek verzeichnet diese Publikation in der Deutschen National-bibliografie; detaillierte bibliografische Daten sind im Internet über http://dnb.d-nb.de/ abrufbar.

Imprint:

Copyright © 2008 GRIN Verlag GmbH
Druck und Bindung: Books on Demand GmbH, Norderstedt Germany
ISBN: 978-3-640-12502-9

This book at GRIN:

http://www.grin.com/en/e-book/112847/virtualization-101-for-small-business-and-enterprise

June 3, 2008

Virtualization 101

for Small Business and Enterprise

Richard Honkanen

<u>Table of Contents</u>

**Virtualization 101
for Small Business and Enterprise**

<u>Executive Summary</u>

<u>Background and Purpose</u>

Virtualization is a new and emerging technology that is rapidly growing, taking industry by storm. In business, virtualization has become a new buzz word, but not many decision-makers actually understand what it is. In today's fiercely competitive world, it is crucial for decision-makers to learn more about virtualization to see if it can give their organizations leverage over their competition. Not only that, virtualization can save businesses of all sizes a lot of money.

<u>Findings and Recommendations</u>

Summed up in one sentence, virtualization is an information technology (IT) logistics strategy that can save businesses time and money. Today, most IT data centers use one machine for one server (e.g. one machine to serve e-mail, one for data, one for the website, etc.), causing server sprawl. Eventually businesses run out of room for new machines, leaving them stuck in a bad position. With virtualization, each server's programs, files, and processes are merged into one file (an image), and is then placed among other images on one machine. Consolidation of IT resources improves server utilization and consequently increases efficiency, lowers the electricity bill, and reduces physical footprint. It also makes deploying new servers cheap. A different approach to virtualization is making multiple physical resources *appear* as one physical resource, simplifying data-center operations and network administration through one consolidated virtual environment.

The advantages of virtualization are almost infinite, but the main points are business continuity, saving money through outsourcing data-processing (cloud computing), and preserving legacy applications. The disadvantages are possible security problems, an initial investment up-front, and uncertainty about the roll-out process (e.g. bugs that take time to work out).

<u>Bottom Line</u>

For some organizations, the uncertainty and cash involved may not be worth the troubles and headaches it may experience down the line. However, enterprises cannot afford *not* to look into virtualization seriously if they wish to remain competitive. Chris Wolf, a senior analyst at Burton Group, "bets the administrators operating in the IT trenches see the cost of virtualization as easily justifiable." Ultimately, the onus is on business decision-makers to decide whether virtualization is a technology that would benefit their organization; however, the bottom line is they need to make an educated decision on this matter.

Virtualization 101
for Small Business and Enterprise

Introduction

Background and Purpose

Virtualization is a new and emerging technology that is rapidly growing, taking industry by storm. In business, virtualization has become a new buzz word, but not many decision-makers actually understand what it is. Their knowledge is limited at best, or they have simply dismissed it based on one magazine article. Reports available today either do not give enough information, give too much technical information, or are written by a vendor and consequently biased towards their offerings. Most reports only cover one or two aspects of this vast new platform for business computing, doing it no justice.

In today's fiercely competitive world, it is crucial for decision-makers to learn about virtualization to see if it can give their organizations leverage over their competition. Not only that, virtualization can save businesses of all sizes a lot of money. This report brings together information from everywhere into one nice package that will leave readers with everything they need to know about the matter. After reading this report, readers will be able to see in which situations virtualization is a cost-effective and worthwhile endeavour for small business and enterprise to undertake.

Scope and Limitations

This report will look at whether server virtualization is useful and worthwhile for businesses to implement. Specifically, it will cover in detail certain situations and applications where virtualization may or may not give businesses leverage. This report is not exhaustive and will not go into technical details. It will be written in plain language so that managers and executives can easily digest the information while still preserving all important terminology.

For the purpose of this report, a small business is defined as having 5-100 employees and annual revenues of $10,000 to $10,000,000, while an enterprise is a multi-million dollar global operation consisting of over 500 employees and revenues of over $10,000,000.

Methodology

The information in this report has been gathered from reputable online sources, including information technology (IT) magazines, business magazines, vendor websites, IT research groups, and technical whitepapers.

<u>Thesis</u>

Virtualization is a new and emerging technology that has become a buzz-word in the business world. With promises of higher efficiency, productivity, and lower operating costs, big players such as Microsoft and IBM are getting in on the action. This report sets to find out whether it is really all it has been made out to be.

Report Organization

This report begins with a primer on what virtualization is. It then proceeds to answer the big questions that every decision maker should know, followed by detailed analysis of situations in which virtualization can be applied. Security concerns are addressed next, and then a summary of pros and cons is presented before the conclusion. In the appendices, readers can find a simple illustration of virtualization and a review of two free virtualization applications.

Virtualization 101

<u>What is Virtualization?</u>

Virtualization, dating back to the 1960s, is a technology that allows one physical computer to do multiple tasks. In its current form, circa 1999, virtualization exploits under-utilized physical resources by merging multiple physical servers into one box. The beauty of this technology is that each virtual server, or virtual machine (VM), can run an entirely different operating system than the physical computer (the host operating system [OS]). By using physical resources more efficiently, small business and enterprise alike can save a considerable sum of money.

A different approach to virtualization is making multiple physical resources *appear* as one physical resource, simplifying data-center operations and network administration through one consolidated virtual environment. In both cases, the key is to achieve higher efficiency and productivity, save money, and simplify operations.

<u>Who is Virtualization for? Why?</u>

Virtualization is for businesses that have a considerable amount of servers and IT resources. It may not be of much use to businesses with only one or two servers, but it can prove useful if they need to purchase a new machine that will not support their legacy applications.

For small business and enterprise, a common problem is server sprawl. An organization usually realizes it exhibits the symptoms of server sprawl when it runs out of space in the server room and its electricity bill is sky high. There are countless servers, each dedicated to only one purpose, all requiring time and money to be updated, protected, maintained, managed, and cooled. With virtualization, a company can drastically reduce the number of physical servers it needs, reduce its electricity bill, and focus its IT efforts on other endeavours.

Virtualization is beneficial, as Yoshihiko Oguchi and Tetsu Yamamoto have aptly stated in the January 2008 edition of the *Fujitsu Scientific & Technical Journal*:

> IT systems have become increasingly larger and more complex, thus making it more difficult to build an optimal IT infrastructure in today's rapidly changing business environment. Server virtualization represents a base technology for addressing this problem. It enables the flexible construction of virtual servers with almost no hardware limitations, and consequently reduces the total cost of ownership (TCO) and makes it easier to use virtual servers in the changing business environment. (46)

In essence, virtualization is an IT logistics strategy that can save time and money.

The Four Pillars of Virtualization

At the core of virtualization, VMware, the company that created contemporary virtualization, lists four key benefits that drive the quick adoption of this technology ("Virtual Machine"). These benefits are what this author has christened *the four pillars of virtualization*: compatibility, isolation, encapsulation, and hardware independence.

Compatibility

Each VM runs its own OS (termed "guest OS") and virtualizes industry-standard hardware. This allows users to run programs normally within the virtualized environment seamlessly. Likewise, the guest OS is able to run on the host OS by way of a hypervisor, a program that manages VMs. Users simply launch the hypervisor program the same way they would open up a word processor or an internet browser, and then select which OS they would like to run.

Isolation

Provided the host computer is powerful enough to run multiple VMs at the same time, every VM that is running is completely isolated from the others. If one of them crashes, the remaining VMs are not affected; they are able to run without conflict. In a traditional computing environment, if a program crashes, it has the potential to cause the whole computer to crash, requiring a reboot. This feature of VMs is why they are perfectly tailored for mission-critical enterprise applications.

Encapsulation

While the hypervisor must be installed on the host OS, each virtual environment has its own independent file. For example, a Windows 98 VM image with accounting software and numerous client files is seen as only one file by the host OS. This allows for easier management and portability of VMs.

<u>Hardware Independence</u>

Like a picture file, each virtual image can be easily copied and transported between different machines without breaking the virtual environment. In the above example, the image file can be transferred from a Linux machine to a Macintosh computer, and the virtual Windows 98 environment, with its accounting program still installed and all the client files intact, would run uneventful on the Macintosh as if it had never moved. The portability and easy storage/backup design of VM images makes it simple for enterprise to deploy and retire VM images on-the-fly.

Common Questions

<u>Is virtualization a fad?</u>

NO! Virtualization is a new platform and strategy that enables businesses to save money and make their IT resources work more efficiently. Further, it removes the boundaries that traditional networking of traditional server infrastructure entails. Virtualization, in effect, is a stepping stone to achieve cloud computing. Refer to the Applications of Virtualization section for more details.

<u>Is virtualization more efficient, productive, and cost-effective in the long run?</u>

Virtualization is definitely all of the above. By consolidating an organization's IT resources, server utilization goes up. This in turn delivers immediate and permanent cost savings (electricity, new servers, administration, etc.). Productivity of work teams also goes up because server procurement is no longer a long and expensive hassle. Refer to the section entitled Virtual Infrastructure under Applications of Virtualization for more information.

<u>Does virtualization replace the need for small business to have tech-support on payroll?</u>

Yes and no. Virtualization reduces administration time, so it may reduce the number of hours that a small business needs a network administrator. However, it does not replace the need for someone to manage the computers, whether they are physical or virtual.

<u>Are the cost-savings for enterprise negligible?</u>

No. In fact, at the enterprise level, the cost-savings are *astronomical!* According to VMware's Total Cost of Ownership (TCO) calculator, infrastructure virtualization for an enterprise in the financial services industry can deliver savings of over $4.7 million dollars, representing a 761.7% return-on-investment (ROI). It even claims that the payback period to realize savings is as short as four months!

Research done by Wafa Moussavi-Amin, analyst and general manager of IDC Europe, an IT market intelligence group, supports VMware's claim: "[...] companies around the globe wasted over $140 billion dollars in the past year for unused server capacity" ("Virtualization Results").

<u>Where is virtualization at today, and is it useful in its current form?</u>

Virtualization is no longer in its infancy, but is still in the growth stage. It is also no longer being used for just testing and development. Andreas M. Antonopoulos, senior vice president & founding partner of Nemertes Research, has found that "[...] among those who have adopted virtualization technologies, up to a quarter are deploying it on production systems."

While virtualization technologies are still being cultivated, it can achieve endless possibilities for enterprise today by removing the limitations of traditional IT infrastructure architecture.

Applications of Virtualization

In an article entitled "Beyond Server Consolidation," Werner Vogels, chief technical officer (CTO) and vice-president of Amazon.com, states that "today the main application of virtualization technology in the enterprise is to combat server sprawl through virtualization-based consolidation" (1). After talking with CTOs and CIOs (Chief Information Officers) around the world, he estimates server utilization to be at a dismal 5 to 12 percent. "With more powerful servers entering the data center every day, the utilization number is decreasing rather than going up" (1). However, server virtualization is not the only application of this new technology.

<u>Business Continuity and Disaster Recovery</u>

In industry, the general concept of backups and disaster recovery (DR) is referred to as business continuity and availability (BC&A). When a server goes down, companies lose money and potential clients to the competition. However with virtualization, the new darling technology can save businesses from downtime and lost customers.

Chris Wolf, a senior analyst at Burton Group, explains that with virtualization, disaster recovery can be completed in a fraction of the time it traditionally takes. He adds that he has "known several organizations who [sic] were never able to complete a successful DR test on physical servers. With virtualization, DR processes can be validated in hours."

Taking it one step further, Kevin Epstein, former director of VMware, says the whole DR process can be automated! In *Linux Journal*, he illustrates this concept with the example of a rack of servers going up "in flames." With a virtualized system, the jobs of the burning servers can be routed to physical servers with available resources without any human intervention (similar to the load-balancing concept detailed below).

<u>Cloud Computing</u>

One major roadblock enterprises experience today is proper resource management. Physical servers often take months to acquire and deploy, and cost the company thousands of dollars in setup and training expenses, only to sit in a corner gathering dust after a short period of time. However, just as one can easily use a 10-10 code to dial long-distance without subscribing to a monthly plan, small business and enterprise alike can tap into powerful mainframe computers through the internet to do intensive number crunching. By creating a special application for a specific task inside of a virtual machine running on a computer in their office, it can communicate with a super-computer somewhere else in the world to do intensive, computational work for a one-time special project.

Werner Vogels asserts that "[...] acquiring and releasing resources based on demand is becoming an essential strategic tool" (2). Deploying virtual servers on an as-needed basis not only saves time and money, but has also turned into a lucrative niche market: "[...] the pay-as-you-go model of the Amazon infrastructure service," Elastic Compute Cloud (EC2), enables enterprises to change their resource acquisition cycles "from months to minutes" (Vogels 2). "One Department of Defence IT architect reported that the department's software prototype normally would cost $30,000 in server resources, but by building it in virtual machines for Amazon EC2, in the end it consumed only $5 in resources" (Vogels 2). Enterprise also use these "[...] utility computing services to address their needs for overflow and peak capacity; this way they can deal with uncertainty in demand without big investments in hardware that will be idle most of the time" (Vogels 3). Not only does this utility model of cloud computing save time and money, it also allows smaller organizations to compete with large conglomerates on the same level.

<u>Load-Balancing</u>

"An extreme example of the use of virtual machine migration is application parking" (Vogels 2), also known as live-migration technology or load-balancing. Simply put, VMs that are idling or using little resources are placed together on one physical server. However, when one of them starts to do more intensive work, the entire VM is automatically sent to a different server that is idling, all without interruption. Thus, instead of having one server per task, a company can get rid of a few physical servers so long as the VMs will not all concurrently tax their IT resources.

Security Concerns

Consolidation of physical servers into VMs can present its own set of challenges. If not properly inventoried and managed, a business can end up with several hundred different and defunct VMs (Santosus)! It is very easy to copy and create a new VM to contain each updated piece of software for compatibility and rollback purposes, but all of those copies tax a company's limited storage and pocket-book through licensing fees. Also, the fact that not every VM is online creates a security nightmare. If each offline VM is not set to wake up at regular intervals, it cannot be updated with the latest security patches.

To add even more headaches, security measures for hypervisors are weak at best. When hypervisors were designed and implemented by enterprise, the technology never went through careful scrutiny of its security model. Now VM vendors are playing catch-up by releasing security patches for vulnerabilities that were never analysed carefully and existed in a live setting for a long time. If a hypervisor is compromised by a malicious attack, an outsider can gain complete access and full control over a corporation's servers and sensitive data. On the other hand, vendors argue that because a hypervisor contains relatively little code, it is much more difficult to find a security hole. Further, the hypervisor is designed to be invisible to internet traffic and the VM, and to-date there have been no successful attacks on hypervisors.

In contrast, assuming that security of the host OS and physical server is adequate, management and control of sensitive data can dramatically increase through virtualization. By housing all of the data in a central location, IT administrators can set file permissions to deny or allow certain users access to certain files. The real beauty of this setup lies in the method of distribution. By deploying VMs onto computer terminals or thin-clients (computers that rely on the server to do most of the processing), users can be placed in a contained, locked-down environment where sensitive data cannot be copied to their terminal's hard drive. This way if someone steals an executive's laptop, for example, the data is still safe on the servers, away from prying eyes.

Nonetheless, even if there is only one security breach on one server, an enterprise's whole data-center could be fully compromised. Because VM images of basic components such as a server OS would be standardized by IT staff for simplicity's sake, one security vulnerability would mean that every virtual server is vulnerable. If that one vulnerability were to be exposed, a malicious user could access *all* resources because *all* the servers would have that security hole.

Virus and malware protection would also be spotty at enterprise-level. On normal computers, anti-virus and anti-spyware programs are updated almost every day. But VMs do not always remain on for long periods of time, especially if they are deployed and retired often. This means that they would not be able to regularly receive updated security definitions. Plus, IT staff usually do not have the time nor deem it necessary to completely secure and scan them for threats. Coupled with the threat of a single un-patched security hole affecting an entire organization's IT resources, Antonopoulos of Nemertes Research sums up the dangers well: "Standardization of images is like dry tinder to a fire: a single piece of malware can become a firestorm engulfing the entire pool of servers."

To add more complexity to the matter, load-balancing VMs automatically poses a new issue: VMs that move around, receiving new internet protocol (IP) and Media Access Control (MAC) addresses with each move, are difficult to firewall. Traditional hardware and software firewalls are unable to protect VMs that constantly move around in an enterprise setting. As with VMs that get deployed and taken down quickly, their ports may not all be secured well, leaving them vulnerable to a malicious attack.

While virtualization can revolutionize a traditional enterprise's data-centre and offer a world of new possibilities, the new security risks that come with it are 1:1. This author highly recommends that people involved with implementing virtualization at a business thoroughly question their vendor/sales person/consultant about security, and consider the implications and cost of compromised data.

Pros and Cons of Virtualization

<u>Pros</u>:

- Business continuity and availability
- Business agility and scalability of software
- Flexible infrastructure
- Server consolidation
- Improve efficiency of IT resources
- Greater control of sensitive data
- Lower costs of ownership for servers
- Improved end-user security
- Keep both legacy and new applications

<u>Cons</u>:

- Complexity of new infrastructure architecture
- Weak security of hypervisor and standardized VM images
- Administration and training costs
- Hit-and-miss capacity planning
- Virtualization creates cost accounting headaches
- Licensing costs
- Non-standard proprietary servers are difficult to virtualize
- Possibly large initial investment into new hardware and software
- One major hardware failure could mean many virtual server failures
- Scalability of hardware
- IT administrators' responsibilities all become blurred

Conclusion

The decision to pursue virtualization is a huge step that carries risk and costs money. For some organizations, that risk may not be worth the troubles and headaches they may experience down the line. However, for big business like the enterprise, they cannot afford *not* to look into virtualization seriously. In order to stay competitive and have leverage over the competition, organizations must stay with the times. For small business, revolutionizing their IT resources may not result in such a dramatic cost savings or efficiency difference. As the old adage goes, "if it ain't broke, don't fix it." This author believes that organizations of all types and sizes should at least look into the possibilities that virtualization can bring. Chris Wolf "bets the administrators operating in the IT trenches see the cost of virtualization as easily justifiable." Ultimately, the onus is on business decision-makers to realize by the end of this document whether virtualization is a technology that would benefit their organization; however, the bottom line is they need to make an educated decision on this matter.

<u>Works Cited</u>

Antonopoulos, Andreas M. "Virtualization Risk Analysis: A risk analysis of large-scaled and dynamic virtual server environments." <u>The Nemertes Research Group Inc.</u> 30 May 2008 <http://www.nemertes.com/issue_papers/virtualization_risk_analysis>.

Epstein, Kevin. "Virtualization 2.0: Where the Sidewalk Ends." <u>Linux Journal</u>. 1 Feb. 2008 <http://www.linuxjournal.com/article/9954>. 31 May 2008.

Oguchi, Yoshihiko, and Tetsu Yamamoto. "Server Virtualization Technology and Its Latest Trends." <u>Fujitsu Scientific and Technical Journal</u> 44,1 (January 2008): 46, 48. <http://www.fujitsu.com/downloads/MAG/vol44-1/paper06.pdf>. 28 May 2008.

S&T System Integration & Technology Distribution AG. <u>S&T: Virtualization results in significant cost benefits</u>. 11 Mar. 2008 <http://www.snt-world.com/Content.Node/news/pressroom/pressreleases/28047.en.php>. 27 May 2008.

Santosus, Megan. "Virtual Machines Bleed Money." <u>Server Virtualization Blog</u>. 23 May 2008. <http://servervirtualization.blogs.techtarget.com/2008/05/23/virtual-machines-bleed-money>. 26 May 2008.

VMware, Inc. <u>Virtual Machine</u>. 22 May 2008 <http://www.vmware.com/technology/virtual-machine.html>.

---. <u>VMware TCO/ROI Calculator</u>. 23 May 2008 <http://www.vmware.com/products/vi/calculator.html>.

Vogels, Werner. "Beyond Server Consolidation." <u>ACM Queue</u> 6,1 (January/February 2008): 1-3. <http://www.acmqueue.com/modules.php?name=Content&pa=showpage&pid=528>. 29 May 2008.

Wolf, Chris. "Benefits of Virtualization in the Data Center? Priceless!" <u>Data Center Strategies Blog</u>. 12 May 2007 <http://dcsblog.burtongroup.com/data_center_strategies/2007/05/benefits_of_vir.html>. 27 May 2008.